ENGINEERING FRAMEWORK FOR RISK-BASED AUTOCLAVE STEAM STERILIZATION CYCLE DESIGN

A GMP Validation Methodology for Complex Mixed Loads in ADC Biopharmaceutical Manufacturing

Prasantha Laxman Pujari

notionpress.com

INDIA • SINGAPORE • MALAYSIA

ISBN 979-8-90510-159-5

Contents

Abstract

This methodology presents a novel engineering risk-based framework for the design of autoclave steam-sterilization cycles, developed by the author on the basis of thermodynamic mapping integrated with Edge-of-Failure analysis. The principal objective is to establish an original, reproducible, and scientifically substantiated algorithm for the design and validation of saturated-steam sterilization cycles capable of ensuring a sterility assurance level no worse than 10^{-6}. The core scientific contribution of the work lies in moving beyond traditional empirical or template-based validation practices toward a data-driven methodology that systematically links process physics, risk assessment, and experimental boundary testing. In this way, the proposed framework addresses the critical industry deficit of standardized approaches for complex loads while aligning with the increasingly stringent expectations of Annex 1 to EU GMP and ISO 17665-1:2024. The relevance of this development is justified by the identified deficit of standardized approaches to cycle design for complex mixed loads, where traditional template regimes of 121.1 °C/15 min and purely formal validation through testing exhibit insufficient sensitivity to latent failures. The scientific novelty of the methodology lies in integrating precise P/T mapping, a modified FMEA, the master-load concept, individual thermodynamic mapping of worst-case locations using wireless loggers, and aggressive edge-of-failure studies, followed by rigorous correlation of calculated F_0 lethality with biological indicator results. It is demonstrated that the critical determinants of success are the completeness of removal of non-condensable gases and the controllability of pre-vacuum conditioning, rather than the duration of exposure. The methodology is intended for validation engineers, quality professionals, and process engineers in biopharmaceutical manufacturing who are developing modern contamination control strategies.

Keywords: steam sterilization, autoclave, saturated steam, thermodynamic mapping, F_0 lethality, edge-of-failure analysis, FMEA

Introduction

The evolution of regulatory requirements for the manufacture of sterile medicinal products reflects a fundamental shift in regulatory authorities' expectations (Elmadhoun et al., 2025). For many decades, the industry relied on legacy standards and template autoclave cycle parameters, the most widely known of which is an exposure at 121.1 degrees Celsius for fifteen minutes (Rutala et al., 2023). This legacy empirical approach, although historically accepted as providing a basic level of sterility assurance, is now increasingly recognized as scientifically inadequate under contemporary GMP expectations, as it relies on inherited cycle templates rather than on a mechanistic understanding of steam penetration, air removal, and load-specific thermodynamic behavior. Regulatory authorities, including the United States Food and Drug Administration (FDA) and the European Medicines Agency (EMA), no longer consider the documentary evidence of a single validation run sufficient (Jain et al., 2024).

As of 2024, manufacturers are asked to apply the science-based evidence approach in the amended Annex 1 of EU GMP guideline and ISO 17665-1:2024 standard. This concept assumes a deep understanding of process physics, the thermodynamic behavior of equipment, and the implementation of quality risk management principles (European Commission, 2022). Saturated steam sterilization is now viewed as a critical and inherently integrated barrier within the facility's overarching contamination control strategy. Such a strategy requires a continuous life-cycle of control, beginning with initial process design and extending through routine monitoring and preventive maintenance (Galiën et al., 2023).

Despite clearly articulated regulatory expectations, in-depth analysis of existing industrial practices reveals a critical deficit of standardized, data-driven methodologies for designing sterilization cycles for complex mixed loads. Current performance protocols often treat the autoclave load as a static thermal mass and disregard the complex hydrodynamics of the steam–air mixture (Feurhuber et al., 2023). The use of rigid, inherited parameters without consideration of equipment-specific geometry, packing density, the presence of long, narrow lumens, and porous materials can lead to so-called hidden failures. These include the formation of persistent air pockets inside instrument lumens, unpredictable migration of temperature lag zones, and the occurrence of wet packs that compromise the sterile barrier. The traditional reliance on temperature sensors alone, without correlating their readings with the pressure profile, masks thermodynamic anomalies and creates the illusion of successful sterilization when actual microbiological inactivation is not achieved (Feurhuber et al., 2023). The described methodology aims to close this gap by proposing a structured, reproducible engineering algorithm.

The primary aim of the present methodology is to establish a reproducible, scientifically grounded algorithm for the development and validation of steam sterilization cycles that unconditionally delivers a sterility assurance level not worse than 10^{-6}. This objective is achieved through the integration of high-precision thermodynamic mapping, rigorous physical calculations of accumulated lethality, and verification of process robustness by conducting edge-of-failure studies. The proposed approach transforms validation from a formal, retrospective testing procedure into a proactive engineering design tool that ensures an exceptional level of patient protection and full transparency for regulatory audits.

CHAPTER 1

Theoretical Framework: Thermodynamics & Risk Assessment

1.1. Bridging Thermodynamic Theory with Practical Validation Decisions

The design of a reliable steam sterilization process is grounded in fundamental laws of thermodynamics that link theoretical physics of heat and mass transfer with practical decisions made during validation. The key agent of microbial inactivation is saturated steam. Its effectiveness is driven by the transfer of latent heat of vaporization. At the moment of contact between saturated steam and the colder surface of the load, an instantaneous phase transition from the gaseous to the liquid state occurs. This transition is accompanied by the release of substantial thermal energy, which is directly transferred to the material and induces rapid coagulation and denaturation of microbial cell proteins (Pletzer et al., 2025). In addition, steam condensation causes a sharp local volume reduction, generating a low-pressure zone that promotes the further entrainment of fresh steam into the deepest cavities and capillaries of the load.

Another important feature of thermodynamic control is the distinction between saturated, superheated, and steam-air mixtures; superheated steam is at a higher temperature than the saturation temperature at the same pressure and can be treated as a dry gas thermodynamically. Its heat transfer coefficient is orders of magnitude lower than that of saturated steam condensing. As a consequence, it cannot efficiently deliver thermal energy to microorganisms, thereby drastically reducing the microbicidal efficacy of the process

(Baker et al., 2025). The presence of non-condensable gases, such as residual atmospheric air, oxygen, or carbon dioxide dissolved in feedwater, constitutes a physical barrier to sterilization. Air is an excellent thermal insulator. Air pockets that form inside tubular components, filters, or densely packed materials prevent direct contact of steam with the surface, which is a primary mechanism for the emergence of hidden defects.

A central scientific innovation of the methodology is the author's conversion of classical thermodynamic principles into a practical, high-resolution validation tool capable of revealing hidden sterilization defects. Rather than treating pressure and temperature as routine monitoring variables, the framework redefines their correlation as the most sensitive and scientifically defensible indicator of effective air removal. This breakthrough is based on Dalton's law of partial pressures, which states that the total chamber pressure equals the sum of the partial pressures of water vapor and non-condensable gases (Tessarolo & Masè, 2025). Consequently, when residual air persists within a difficult-to-penetrate load, the local steam partial pressure falls below the total pressure recorded by the chamber sensor, exposing the thermodynamic mechanism by which undetected air pockets undermine sterilization efficacy. practical validation. Also, the local saturation temperature in the air pocket will be lower than the intended sterilization temperature, even when the autoclave controller registers attainment of the setpoint. The use of the Antoine equation or standard thermodynamic water and steam tables allows engineers to calculate the theoretical saturation temperature solely from the reading of an independent pressure sensor. A significant negative deviation between the actual temperature recorded by wireless loggers deep within the load and the theoretically calculated temperature provides robust, unambiguous evidence of air presence (Pletzer et al., 2022). Thus, integrating pressure–temperature correlation transforms the detection of residual air from a field of subjective conjecture into a domain of exact engineering calculations that fully comply with the stringent requirements of ISO 17665-1:2024 (ISO, 2024).

1.2. FMEA Methodology for Autoclaves

The transition of the biopharmaceutical industry toward a quality-by-design paradigm necessitates integrating quantitative and qualitative risk analysis tools directly into autoclave cycle development. The modified failure modes and effects analysis methodology is specifically adapted to the physical specifics of sterilization equipment operation and links each process parameter with a concrete mechanism of potential failure. Instead of dealing with abstract or purely procedural risks, this approach converts uncertainty into testable hypotheses and focuses on critical process parameters such as the depth of vacuum pulses, the temperature ramp rate, and the equilibration time.

The methodology involves a systematic decomposition of the steam sterilization process into clearly defined phases, including air removal, heating, sterilization plateau, and pulsed drying. At each stage, an expert team identifies potential failure modes, analyzes their root causes, and develops preventive engineering controls. Risks are evaluated using a standard matrix that considers the severity of consequences for product sterility, the probability of failure, and the ability of existing monitoring systems to detect the failure before the product leaves the autoclave. In the context of sterilizer validation, particular attention is paid to failures that result in inadequate or delayed heat penetration, as such defects cannot be detected by standard sensors installed in the chamber drain line.

To systematize this analytical approach and embed it directly into validation practice, a detailed risk matrix is developed that goes beyond a static summary table. In the proposed methodology, it functions as a dynamic analytical tool that translates abstract thermodynamic and operational uncertainties into structured, testable engineering hypotheses. Through this function, the matrix actively guides cycle development, sensor placement, parameter optimization, and the selection of targeted mitigation strategies. Table 1 therefore presents the core analytical framework for identifying, prioritizing, and controlling vulnerabilities associated with critical steam-sterilization cycle parameters.

Table 1. FMEA risk matrix for autoclaving cycle parameters

Autoclave cycle stage	Critical Process Parameter (CPP)	Potential failure mode & mechanism	Thermodynamic root cause	Engineering controls & mitigation measures
Dynamic conditioning	Rate and depth of vacuum pulses	Formation of a stable air pocket in long lumens	Insufficient pressure gradient to overcome the aerodynamic resistance of the tubing	Optimize pulse profile, reduce pressure change ramp rate, and add a vacuum hold phase
Heating phase	Temperature ramp rate	Thermodynamic instability and temperature variability	Different heat capacities of mixed loads lead to uneven condensation	Use modulated steam injection, strict pressure control, and increase heating time
Sterilization plateau	Temperature equilibration time	Cold spot formation; failure to reach required lethality	Presence of non-condensable gases that impede convective heat transfer	Implement a strict equilibration criterion (< 15 seconds), control pressure–temperature correlation
Post-conditioning	Pulsed vacuum drying time	Wet packs; risk of product recontamination	Insufficient time or vacuum depth to evaporate residual condensate from porous materials	Optimize item orientation in the chamber, use plastic baskets, and extend the drying phase duration

Implementing this risk matrix enables validation engineers to embed variability-compensating mechanisms directly into the software architecture of the autoclave. As a result, the autoclave ceases to function as a black box and becomes a fully transparent process in which cause–and–effect relationships between the physics of the medium, equipment settings, and sterility assurance metrics are amenable to strict control and mathematical justification.

1.3. The Master Variable Load Concept

The concept of a master load is one of the central pillars of the validation program and provides a scientifically defensible basis for demonstrating process scalability. The author's original contribution to this concept lies in its quantitative definition: the master load is not selected intuitively as merely the heaviest or most cumbersome configuration but is established through a structured engineering analysis of the load characteristics that create the highest thermodynamic resistance to effective steam penetration and heat transfer. In this way, the master load becomes a calculated worst-case model, designed to generate an indisputable body of evidence that the sterilization process remains robust under the most demanding and physically unfavorable conditions. The principal aim is to obviate the need to validate every equipment configuration used in production. A single worst-case scenario is built where all the factors that are most unfavorable for proper steam sterilization thermodynamics are contained in either a single hypothetical, or a real, validation configuration. If the reliability of the cycle is confirmed on such a maximal load, it gives the regulators scientific certainty that any production configurations will be reliably sterilized if they are less massive or less geometrically complex.

The first step in identifying the worst case is to analyze each of three key physical parameters: the maximum thermal mass of the load. It is determined by the mass of heavy-metal components (pump housings, steel manifolds and large valves) which have a large thermal inertia and take a large amount of latent heat to heat them up. This leads to a longer time to reach operating temperature and increased risk of a condensate plug. The second, and often limiting, factor is the poorest steam penetration achieved with long silicone tubing with narrow internal lumens and multilayer hydrophobic filters.Another characteristic of these structures is the existence of a dead-end effect, in which gas diffusion is low, but where gas trapping is more likely. Finally, maximum configuration density is a third factor. Objects packed in baskets or multilayered textile wrapping impede the free flow of steam, converge convection currents, and thus delay equilibration of the temperature field in the autoclave chamber.

To organize the selection of the master-load elements, a decision tree including all physical parameters and validation methods is provided in Table 2.

Table 2. Logical structure of master load formation

Load physical parameter	**Worst-case indicator**	**Thermodynamic rationale for vulnerability**	**Verification & mapping method**
Extreme thermal inertia	Solid metal manifold with mass > 4 kg	Slow heating dynamics require an extended time to achieve thermal equilibrium between the surface and the core	Place a temperature data logger at the geometric center of the metal mass
Increased aerodynamic resistance	Multi-layer porous materials and process garments	Fibrous structure effectively traps cold air, creating a barrier to saturated steam penetration and condensation	Place wireless temperature sensors and steam chemical integrators in the very center of the folded porous pack
Complex internal geometry	Silicone hoses up to 2 m long with minimal inner diameter	Aerodynamic dead-end effect prevents air removal during pre-vacuum; risk of condensate plug formation	Place a temperature sensor and a biological indicator at the central point of the tube, with an end sealing
Maximum configuration density	Tightly packed metal parts in closed sterilization boxes	Critical reduction of open surface area available for direct contact with the steam environment	Precision thermometry in the gaps between tightly adjacent packs inside the basket

The master-load approach not only reduces the number of qualifications of individual items but also determines a tool that can be easily transferred to other equipment units and sites, enabling a way to show the functional robustness of the cycle against extreme loads in the design space.

1.4. Selection of Sterilization Approach: Overkill vs. Bioburden

The choice of how to calculate sterilization lethality is key to the design of the sterilization cycle, how it is monitored, and the supporting evidence that will be presented for auditing and regulatory review. There are two fundamentally different, but equally accepted, approaches to this issue in world biopharmaceutical practice. These two approaches are the product bioburden-based approach and the overkill approach (Ben-David, 2022). Understanding the physical, microbiological and logistical limitations of both approaches is important to the establishment and validation of contamination control.

With the bioburden-based method the quantitative and qualitative knowledge of the pre-sterilization microflora of the product combined with the knowledge of the resistance to thermal treatment expressed as D-value may be used as information by the engineer to achieve minimally aggressive short cycle times in steam sterilization. This process is particularly useful for thermosensitive protein solutions, complex dosage forms, or low-melting polymeric containers that would deform or dissolve when exposed to heat The theoretical calculation of the ideal cycle duration in this approach requires logarithmic transformation of the initial number of colony-forming units (CFU), addition of the target logarithmic reduction factor (typically six logs to ensure SAL 10^{-6}), and multiplication of the resulting sum by the D-value of the most resistant microorganism identified in the production environment.

The practical implementation of this approach requires substantial analytical effort from quality control laboratories. Continuous monitoring of the microbiological background is required because any seasonal shift, change in raw material suppliers, or emergence of a more resistant microorganism can render a previously validated gentle cycle completely ineffective (Ramos et al., 2024).

The overkill method represents the conceptual counterpart to the bioburden-based approach. It assumes an extreme and highly unlikely worst-case scenario. In this scenario, the load surface is contaminated with an exceptionally high concentration of thermoresistant bacterial spores. Geobacillus stearothermophilus spores are traditionally used as the reference. The method requires conditions sufficient to achieve at least a 12-log reduction in the population of this highly resistant reference microorganism, which creates a substantial process safety margin. For the sterilization of metallic equipment, silicone tubing, filters, and mixed loads that can physically withstand aggressive thermal exposure, the overkill method is an unequivocal priority. This approach eliminates uncertainty associated with fluctuations in natural bioburden, removes the need for routine typing of resident microflora, and aligns seamlessly with the robust engineering concepts expected by the FDA and EMA.

A comparative characterization of these fundamental approaches is provided in Table 3 to justify the selection of the cycle architecture.

Table 3. Comparative analysis of sterilization strategies

Parameter for comparing approaches	Overkill Approach	Bioburden Approach
Target Sterility Assurance Level (SAL)	At least 10^{-6}, calculated based on reference thermoresistant spores	At least 10^{-6}, calculated based on the natural resident microflora
Required log reduction	Minimum 12-log reduction for the reference microorganism	Case-by-case calculation, typically 6–8 log reduction
Regulatory & validation focus	Exclusive confirmation of uniform heat distribution and complete steam penetration	Strict, costly, and continuous monitoring of the load's microbiological cleanliness prior to sterilization
Technical applicability to mixed loads	Very high; conservatively compensates for bioburden variability across heterogeneous materials	Low; requires overly complex, multi-factor control for multi-component or unpredictable loads
Positioning in the proposed methodology	Primary and non-alternative approach for complex geometries; forms the basis for F_0 calculations	Strictly limited use; permitted only for product components with critical thermal sensitivity

Within the presented methodology, the overkill method is selected as the fundamental basis. It ensures that, even in situations with unforeseen aerodynamic resistance or transient temperature gradients within the complex cavities of the master load, the final accumulated lethality remains within a safe range. This outcome aligns directly with the requirements for holistic contamination control.

CHAPTER 2

Data-Driven Cycle Design (Engineering Phase)

2.1. Defining Phase Parameters (Engineering Runs)

The development of a robust sterilization cycle architecture is grounded in empirical data obtained from a series of engineering runs. These runs serve the double purpose of having a phase parameter optimization tool and a characterization of the hardware thermodynamics as a function of the configuration. The design approach in this chapter is engineering rigorous, and has been both well tested on the GMP autoclaves and has demonstrated robustness beyond the anatomy of a single machine. So the main objective of this stage of the investigation should be to make the parameters of the vacuum pulse as optimal as possible, which only serves to remove atmospheric air from the most diverse load geometries.

By analyzing experimental data, it can be concluded that the hydrodynamics of the steam-air mixture in long silicone lumens is strongly affected by the vacuum level and by the rate of pressure drop in the lumens. Excessively rapid pressure reduction may trigger abrupt boiling of residual moisture, leading to blockage of narrow channels or, more critically, to immediate activation of autoclave safety interlocks associated with saturated steam conditions.

Within the study, four consecutive engineering runs of a full-scale master load were analyzed in detail. During the first two runs, the pressure ramp rate was set optimistically at 30 pounds per square inch per minute. The objective of this setting was to achieve the maximal reduction in total cycle time. This aggressive dynamic regime proved technologically excessive for the given equipment configuration. Both cycles were aborted by the control system due to alarm conditions linked to violations of saturated-steam parameters.

The engineering response to this systemic failure involved a fundamental reconfiguration of the preconditioning logic. The number of vacuum pulses and the pressure differential between peaks were increased, while the rate of pressure rise and fall was markedly reduced. Application of the modified parameters in subsequent runs produced outstanding results. The temperature equilibration time between the chamber drain and the coldest point in the load was less than five seconds. This result provides unequivocal evidence of complete, smooth, and unobstructed air removal from all elements of the complex configuration.

Figure 1 presents the graphical profile of a successfully optimized autoclave cycle.

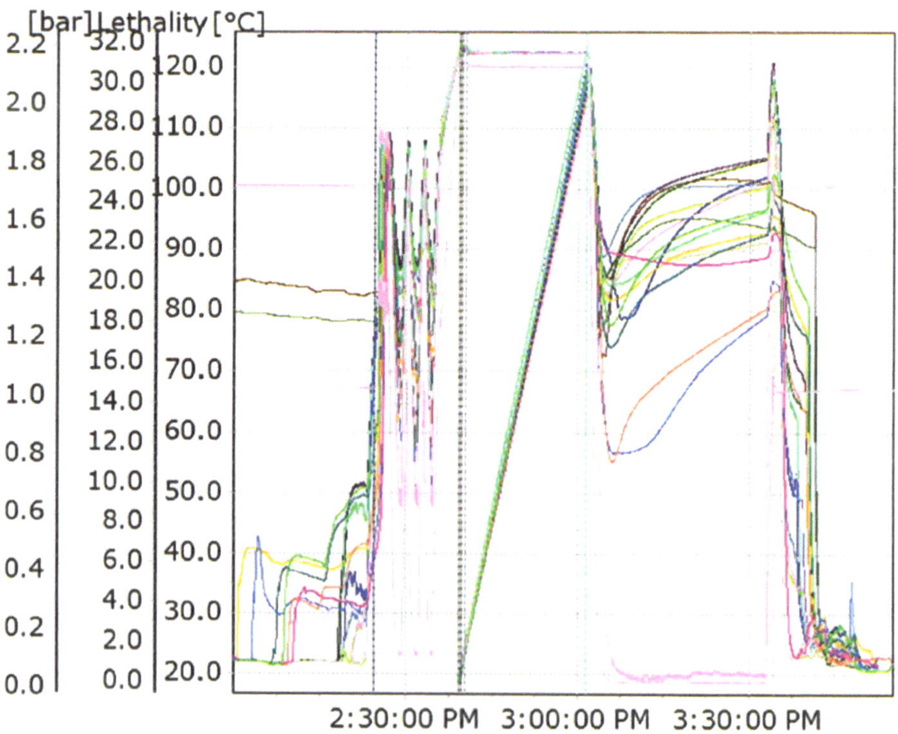

Fig. 1. Graphical profile of a successfully optimized autoclave cycle

The graphical profile in Figure 1 shows a successfully optimized autoclave cycle, recorded in the protocol, that meets all criteria. On the multichannel thermogram, it is clearly visible that after modification of the vacuum phase parameters, all colored curves originating from the numerous independent temperature data loggers converge into a narrow temperature bundle almost immediately after initiation of the heating phase. The exceptional compactness of the phase pattern reflects high process controllability and strong predictability in temperature dynamics. Transitions between deep vacuum, steam injection, the sterilization plateau, and final drying appear graphically sharp, without the prolonged fluctuations and delays that characterize unstable regimes.

2.2. Development Acceptance Criteria

To ensure the most objective and unbiased assessment of engineering runs, acceptance criteria must be established. Within the risk-based methodology, these criteria are designed to exceed minimal regulatory requirements and thereby create a mathematically demonstrable safety buffer.

The fundamental performance metric in this paradigm is the time to temperature equilibration. This parameter is defined as the time interval between the moment the target sterilization temperature is achieved at the reference control point and the moment it is recorded at all measurement points within the complex load. In the context of enhanced risk management, this criterion is interpreted more stringently. It is considered the earliest, most direct, and most sensitive thermodynamic marker of successful air removal. Prolonged or uneven equilibration reliably indicates the presence of local accumulations of non-condensable gases that obstruct convective heat transfer.

Use of this stringent criterion is significantly more effective than simply observing attainment of the setpoint at the chamber controller, because it allows detection of latent steam penetration problems long before evaluation of the process's integral lethality. Figure 2 illustrates the equilibration time.

Equilibration Time

Name:	Equilibration Time
Description:	Time difference from sterilization start (reference temp reaching 121C) and all remaining loggers reaching 121C.

Input parameters

Data Series Type:	Temperature
Data series:	LC 01, LC 02, LC 03, LC 04, LC 05, LC 06, LC 07, LC 08, LC 09, LC 10, LC 11, LC 12, LC 13, LC 14, LC 15, LC 16, LC 17, LC 18
Start Time:	2:39:24 PM
Stop Time:	2:39:39 PM
Output:	Data Series analysis, Global Data analysis
Functions:	Min.

Global Data analysis

Min.	122.4 °C	2:39:24 PM	LC 11

Data Series analysis

ID	Name	Min.	Min. sample time
685941-1	LC 01	122.6 °C	2:39:24
685943-1	LC 02	122.9 °C	2:39:24
685946-1	LC 03	123.5 °C	2:39:24
685948-1	LC 04	123.0 °C	2:39:24
685950-1	LC 05	123.8 °C	2:39:24
686043-1	LC 06	123.2 °C	2:39:24
636557-1	LC 07	123.8 °C	2:39:24
686045-1	LC 08	123.3 °C	2:39:24
686046-1	LC 09	123.2 °C	2:39:24
686049-1	LC 10	122.4 °C	2:39:24
686050-1	LC 11	122.4 °C	2:39:24
686052-1	LC 12	122.4 °C	2:39:24
686056-1	LC 13	122.4 °C	2:39:24
686057-1	LC 14	122.4 °C	2:39:24
686063-1	LC 15	122.4 °C	2:39:24
686067-1	LC 16	122.4 °C	2:39:24
686069-1	LC 17	122.4 °C	2:39:24
686070-1	LC 18	122.5 °C	2:39:24

Fig. 2. Equilibration Time

As shown in Figure 2, a detailed statistical analysis of this critical parameter is documented. Global data analysis shows that, in a properly designed cycle, the target temperature corridor is reached by all sensors almost simultaneously. In validation runs, the equilibration time was less than fifteen seconds, and in the most optimized phases, it was less than five seconds. Such near-ideal synchrony indicates that a coherent temperature field forms almost instantaneously and that the target sterilization window is maintained as a property of the entire thermodynamic system of the autoclave rather than as an isolated characteristic of open regions.

A second acceptance criterion of comparable importance is the load's dryness. Residual moisture after material removal from the autoclave is entirely unacceptable. Water compromises the physical integrity of the packaging's sterile barrier, alters the properties of paper-based filters, and creates a nutrient-rich medium for immediate recontamination of the product upon contact with non-sterile ambient air. The success criterion requires that the increase in the mass of a representative load after completion of the cycle must not exceed one percent of its initial weight. Compliance with this strict parameter is confirmed by a combination of visual inspection for the absence of droplets and precision weighing of the packs. Figure 3 presents the Failed Study Data Autoclave Cycle Graph.

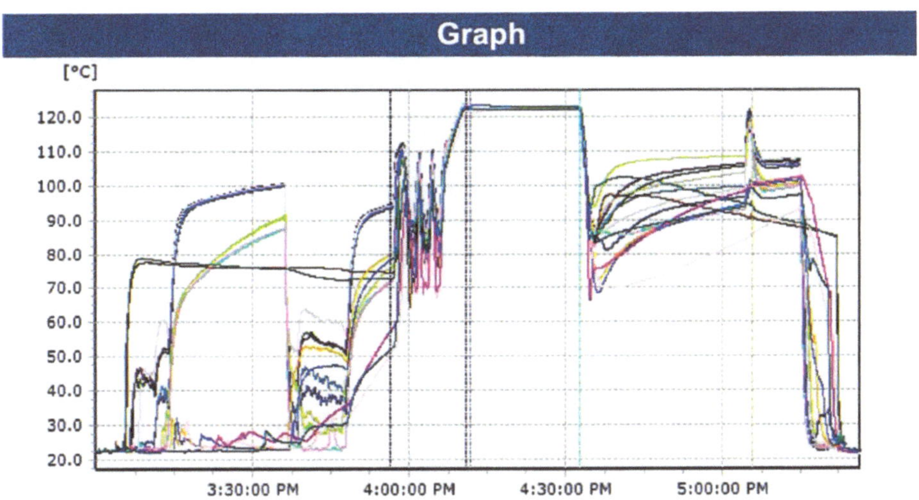

Fig. 3. Failed Study Data Autoclave Cycle Graph

Comparative analysis of thermograms from successful and failed engineering runs clearly reveals the thermodynamic anatomy of failure. The results of the failed study show the temperature curves diverging strongly chaotically in the ramp-up to the operating regime. Having single sensors in hard to reach places can have a strong delay in heating and is dangerous.

In addition, there are continuous up and down variations in temperature during the plateau phase of sterilization, which are a classical physical manifestation of the instability of the steam-air environment, brought about by the expansion and contraction of the air pockets. The sharp visual difference is evidence that strictly maintaining the instantaneous equilibration criterion is directly and tightly correlated to overall process stability and safety.

2.3. Individual Item Mapping

Data on the temperature profiles of the most complex and sensitive load components usually require high-precision metrological instruments and non-standard engineering solutions. Customary large and physically inflexible wired thermocouples usually require breaking the sterility of the packaging or modifying the seal or geometry of a closed container. However, this distortion can change internal aerodynamics and not necessarily provide an accurate image of steam penetration.

For this reason, the methodology's technological foundation is individual thermodynamic mapping of items using miniature, high-accuracy wireless data loggers. These high-precision autonomous devices are placed directly at so-called worst-case locations. Examples include the center of tightly coiled multi-meter silicone hoses, the interior of multilayer porous filters, and positions under valves of complex process equipment. This aggressive risk mitigation strategy reliably prevents unjustified confidence from forming, which inevitably arises when personnel rely solely on readings from built-in autoclave sensors that measure temperature only in the free chamber space.

The logger placement diagram is illustrated in Figure 4.

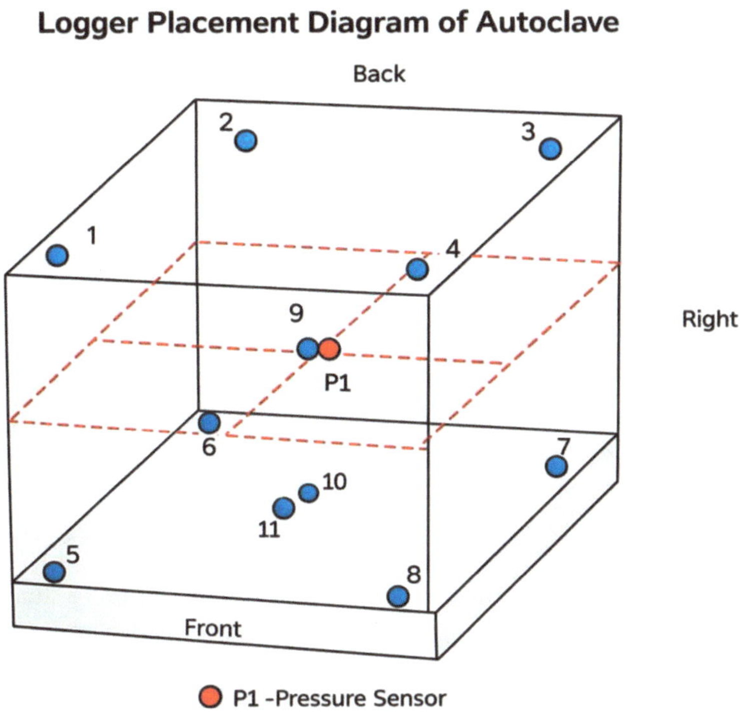

Fig. 4. Logger Placement Diagram of Autoclave

The diagram in Figure 4 provides a detailed illustration of a standardized geometric approach to sensor placement. The working volume of the autoclave chamber is modeled as a transparent three-dimensional parallelepiped with clearly indicated orientation planes. Blue markers indicate temperature loggers strategically distributed across all critical zones. These zones include cold lower corners, side walls, top extremities, and positions inside complex load elements. A red marker designates an independent pressure sensor. It is located in the central part of the chamber to enable continuous conversion of pressure into theoretical saturation temperature, which represents a cornerstone of the methodology. Dense sensor placement enables empirical localization of wandering cold spots, whose actual positions rarely coincide with theoretical predictions from legacy protocols.

The data logger list is shown in Figure 5.

Sensor ID	Name	Validation Equipment	Description	Sample Rate 1/2 (hh:mm:ss)	Type	Sensor group
685941	LC 01	715757	Temperature	00:00:01 / N/A	LC	N/A
685943	LC 02	715739	Temperature	00:00:01 / N/A	LC	N/A
685946	LC 03	715889	Temperature	00:00:01 / N/A	LC	N/A
685948	LC 04	715753	Temperature	00:00:01 / N/A	LC	N/A
685950	LC 05	715913	Temperature	00:00:01 / N/A	LC	N/A
686043	LC 06	715892	Temperature	00:00:01 / N/A	LC	N/A
636557	LC 07	715909	Temperature	00:00:01 / N/A	LC	N/A
686045	LC 08	715920	Temperature	00:00:01 / N/A	LC	N/A
686046	LC 09	715887	Temperature	00:00:01 / N/A	LC	N/A
686049	LC 10	715882	Temperature	00:00:01 / N/A	LC	N/A
686050	LC 11	715915	Temperature	00:00:01 / N/A	LC	N/A
686052	LC 12	715905	Temperature	00:00:01 / N/A	LC	N/A
686056	LC 13	715903	Temperature	00:00:01 / N/A	LC	N/A
686057	LC 14	715902	Temperature	00:00:01 / N/A	LC	N/A
686063	LC 15	715765	Temperature	00:00:01 / N/A	LC	N/A
686067	LC 16	715780	Temperature	00:00:01 / N/A	LC	N/A
686069	LC 17	715799	Temperature	00:00:01 / N/A	LC	N/A
686070	LC 18	715747	Temperature	00:00:01 / N/A	LC	N/A
656186	LP 19	715899	Pressure	00:00:01 / N/A	LP	N/A
N/A	LP 19 - T-Theoretical 1	N/A	N/A	00:00:01 / N/A	TA	N/A

Fig. 5. 18 Ellab Data Logger used and 1 Pressure data logger

Complementing the spatial placement scheme, meticulous documentation of the measurement configuration is critical for developing an audit-ready evidence base. The report confirms the simultaneous use of 18 autonomous temperature loggers with a sampling frequency of 1 second. This frequency is essential for capturing fast transient phenomena during vacuum break. An independent pressure sensor is an absolute requirement. As demonstrated by the analysis, its absence in unsuccessful historical runs deprived engineers of the physical ability to correctly detect insidious residual air (Tessarolo & Masè, 2025).

CHAPTER 3

Validating Robustness: Edge of Failure & Lethality

3.1. Edge of Failure (EoF) Studies

The conceptual apex of the risk-based approach to sterilization process design is the implementation of detailed edge-of-failure studies. Although current international guidance does not include explicit, mandatory requirements for mapping the operating limits of each routine cycle, voluntary incorporation of this aggressive practice into the validation methodology provides compelling evidence to regulators that the process is subject to deliberate, intensive hyper-control (Stewart, 2025).

In line with the principles of ICH Q8(R2), identification of the precise edge of failure enables engineers to thoroughly understand the boundaries of the design space and to quantify how far process parameters may deviate from the optimum before a critical loss of sterility occurs (FDA, 2009).

The methodology for conducting edge-of-failure studies consists of conscious, methodical, and strictly controlled degradation of key cycle parameters. Development engineers intentionally deteriorate sterilization conditions by reducing the duration of high-temperature exposure, decreasing the depth or number of vacuum pulses, or artificially lowering the team generator temperature. Degradation continues until the calculated physical lethality approaches the biological survival threshold for the resistant test microorganisms.

The primary purpose of this destructive testing is to quantify the safety distance between standard operating conditions and the point of inevitable failure.

Within the empirical dataset analyzed for cycle development, the edge-of-failure study was conducted with a specific objective. The objective was to provide unequivocal confirmation that the new autoclave profile remains effective at the point where applied physical lethality, with precision to the tenths of a minute, matches the biological lethality derived from the biological indicator certificate of analysis corresponding to a borderline SAL of 10^{-2}. In the experiment, lethality was carefully adjusted to approximately eleven minutes, and the cycle was then manually terminated by the operator. Subsequent incubation demonstrated that all 15 biological indicators placed in the most thermally inaccessible worst-case locations showed no bacterial growth.

This outcome confirms that the process retains microbicidal efficacy even under artificially extreme depletion. Considering that, in the actual production cycle, the algorithm is configured to deliver at least 30 minutes of accumulated F_0, this edge-of-failure experiment provides documentary proof of a very large mathematical safety margin and the outstanding robustness of the designed system.

3.2. Lethality (F_0) Calculation & BI Correlation

Mathematical modeling of accumulated thermodynamic lethality at empirically identified cold spots constitutes the core of physical qualification for the sterilization process. The F_0 concept is an elegant mathematical tool that enables universal measurement and comparison of the sterilizing capacity of any thermal exposure. This abstraction reduces any real combination of time and fluctuating temperature to an equivalent exposure time at the reference temperature of 121.1 degrees Celsius with a Z-value of exactly 10 degrees Celsius.

The integral equation for calculating accumulated lethality is expressed by the classical formula:

$$F_0 = \int_{t_0}^{t_1} 10 \frac{T(t) - 121.1}{10} dt, \quad (1)$$

where T(t) denotes the actual temperature recorded by the thermosensor at each time point t throughout heating, exposure, and cooling.

This mathematical model provides engineers with exceptional flexibility. It enables the design of effective cycles for biopharmaceutical materials that cannot tolerate aggressive temperature peaks. Required lethality is achieved by proportionally and scientifically increasing exposure time at a reduced temperature. The lethality factor enables accurate regime substitutions. At a process temperature of 118 degrees Celsius, the lethality factor is 10^((118 - 121.1)/10) = 0.489. A lethality of 15 minutes at the reference temperature can be achieved by increasing the process exposure time. This division, 15 minutes/0.489, gives approximately 31 minutes, which is done to ensure that these thermolabile plastics and complex protein suspensions are not adversely affected by the vaporization heat from the sterilization process.

For unequivocal demonstration of sterility to regulatory authorities, physical lethality calculations must be correlated with direct biological inactivation results obtained using indicators inoculated with Geobacillus stearothermophilus spores. The enhanced risk-based methodology mandates placement of a temperature sensor in immediate physical proximity to each biological indicator within the identified cold spot. This practice establishes a direct and scientifically defensible causal link between the recorded temperature profile and the death of test microorganisms.

Summary lethality calculations that illustrate extrapolation to the required SAL are systematized in Table 4. This table presents computational models of lethality for various loading scenarios.

Table 4. Computational models of lethality for different loading scenarios

Calculated thermodynamic scenario type	Process & load input parameters	Intermediate logarithmic calculations	Final engineering cycle-time requirement
Adaptation for extreme bioburden	Initial bioburden: 1,000 CFU; D-value: 1.2 min/log; Target SAL: 10^{-6}	Required log reduction: log(1000) + 6 = 3 + 6 = 9 logs	Ideal minimum cycle time: 9 × 1.2 = 10.8 min at stable 121.1 °C
Temperature equivalence under reduced heating	Engineering target: achieve 15 min accumulated F_0 at actual chamber temperature 118 °C	Lethality factor: 10^((118 − 121.1)/10) = 10^(−0.31) ≈ 0.489	Required exposure at 118 °C: 15 / 0.489 ≈ 31 min continuous hold
Extrapolation to full SAL under temperature shortfall	Current F_0: 22 min; BI population: 8 × 10^5; Minimum recorded temperature: 118.5 °C	Lethality deficit L_a = 22.7 min; factor at 118.5 °C = 10^((118.5 − 121.1)/10) = 10^(−0.26) ≈ 0.550	Required compensatory time: 22.7 / 0.550 ≈ 41.3 min; Total hold time: 82 min

Figure 6 illustrates the Lethality Calculation and the Input Parameter Used.

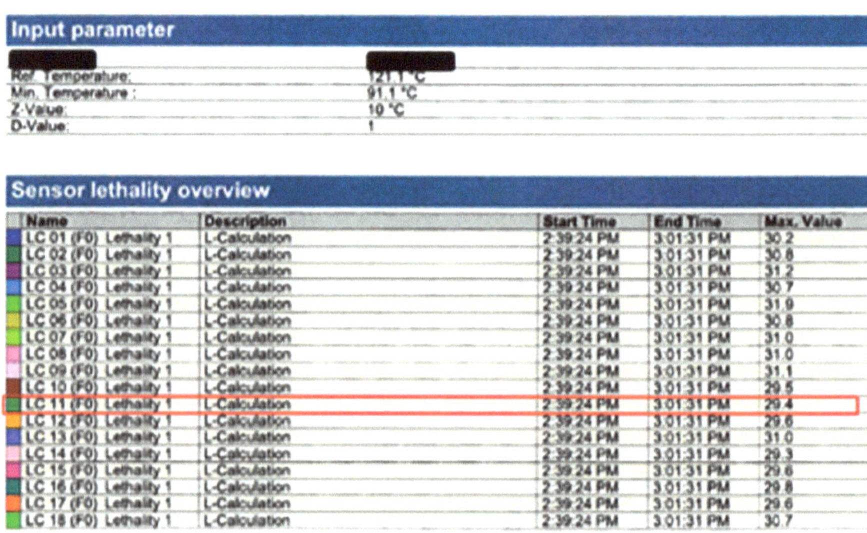

Input parameter

Ref. Temperature:	121.1 °C
Min. Temperature :	91.1 °C
Z-Value:	10 °C
D-Value:	1

Sensor lethality overview

Name	Description	Start Time	End Time	Max. Value
LC 01 (F0) Lethality 1	L-Calculation	2:39:24 PM	3:01:31 PM	30.2
LC 02 (F0) Lethality 1	L-Calculation	2:39:24 PM	3:01:31 PM	30.8
LC 03 (F0) Lethality 1	L-Calculation	2:39:24 PM	3:01:31 PM	31.2
LC 04 (F0) Lethality 1	L-Calculation	2:39:24 PM	3:01:31 PM	30.7
LC 05 (F0) Lethality 1	L-Calculation	2:39:24 PM	3:01:31 PM	31.9
LC 06 (F0) Lethality 1	L-Calculation	2:39:24 PM	3:01:31 PM	30.8
LC 07 (F0) Lethality 1	L-Calculation	2:39:24 PM	3:01:31 PM	31.0
LC 08 (F0) Lethality 1	L-Calculation	2:39:24 PM	3:01:31 PM	31.0
LC 09 (F0) Lethality 1	L-Calculation	2:39:24 PM	3:01:31 PM	31.1
LC 10 (F0) Lethality 1	L-Calculation	2:39:24 PM	3:01:31 PM	29.5
LC 11 (F0) Lethality 1	L-Calculation	2:39:24 PM	3:01:31 PM	29.4
LC 12 (F0) Lethality 1	L-Calculation	2:39:24 PM	3:01:31 PM	29.6
LC 13 (F0) Lethality 1	L-Calculation	2:39:24 PM	3:01:31 PM	31.0
LC 14 (F0) Lethality 1	L-Calculation	2:39:24 PM	3:01:31 PM	29.3
LC 15 (F0) Lethality 1	L-Calculation	2:39:24 PM	3:01:31 PM	29.6
LC 16 (F0) Lethality 1	L-Calculation	2:39:24 PM	3:01:31 PM	29.8
LC 17 (F0) Lethality 1	L-Calculation	2:39:24 PM	3:01:31 PM	29.6
LC 18 (F0) Lethality 1	L-Calculation	2:39:24 PM	3:01:31 PM	30.7

Fig. 6. Lethality Calculation and Input Parameter Used

Validation results confirm the full practical applicability of these calculations. The report includes baseline parameters, including the reference temperature of 121.1 degrees Celsius and the minimum integration start temperature of 91.1 degrees Celsius. The documented maximum accumulated lethality values for the sensors under study ranged from 29 to 32 minutes. This range significantly exceeds the standard industry overkill criterion of 12 minutes.

It is highly significant that even the sensor with the lowest recorded lethality still demonstrated results that guarantee attainment of SAL 10^{-6}. This outcome provides final confirmation of the exceptional robustness of the designed temperature profile.

Despite its mathematical elegance, a critical methodological limitation of the F_0 concept must be fully recognized. The F_0 value continues to increase in the computational system even when a dry hot air mixture is present inside a cavity, and actual biological inactivation of spores is severely impaired or completely halted in the absence of moisture. Therefore, the calculation of integral thermal lethality is considered legitimate and scientifically grounded only after an independent and rigorous pressure–temperature correlation analysis establishes the complete presence of saturated steam throughout the load volume (Miguel et al., 2024).

3.3. Routine Monitoring Strategy

The life cycle of a reliable steam sterilization process does not terminate at the signing of the final validation report. It continues in the critical phase of stringent routine monitoring. This phase provides daily assurance that the process remains within the carefully defined design space and is not drifting due to equipment wear. The advanced routine monitoring strategy tightly integrates simple daily tests with the fundamental precision thermal-mapping results obtained during initial engineering.

The central diagnostic element of daily checks for pre-vacuum autoclaves is the classical Bowie–Dick test. Its primary purpose is to verify the mechanical efficiency of air removal and confirm the absence of micro-leaks in chamber seals (ISO, 2007). Within the modern risk-based paradigm, successful completion of this standardized test alone is no longer regarded as a definitive guarantee of success for complex production loads. It is crucial that preconditioning parameters, including the number and depth of vacuum pulses in the standard Bowie–Dick test cycle, be identical to or rigorously correlated with the corresponding parameters of the designed production cycle.

To avoid dangerous false-positive interpretations of visual results, data from daily chemical tests and critical machine parameters, such as the temperature ramp profile and the equilibration time for each batch, must be continuously subjected to statistical trend analysis. If machine equilibration time begins to deviate systematically from the validated reference, for example, by repeatedly exceeding the strict limit of fifteen seconds, this behavior must be treated as an immediate warning signal. Such drift indicates latent equipment aging, microscopic wear of door gaskets, reduced performance of the vacuum pump, or partial blockage of drain filters.

Implementation of this analytical approach transforms routine monitoring from a monotonous form-filling exercise into a powerful predictive diagnostic tool that integrates seamlessly into the broader preventive maintenance system of a biopharmaceutical facility.

Field Impact & Industry Significance

Field Impact and Industry Significance

Regulatory guidance has created an expectation without a method. EU GMP Annex 1 (2022) and ISO 17665-1:2024 both require science-based sterilization cycle development for complex mixed loads. Neither document tells a validation engineer how to execute it. The result, documented in peer-reviewed literature, is a measurable deficit of standardized, data-driven methodologies one that has produced hidden sterilization failures, unexplained deviations, and compromised sterile barriers across the industry (Feurhuber et al., 2023).

This manual provides the methodology that existing guidance does not.

The framework integrates thermodynamic mapping, pressure-temperature correlation, modified failure mode and effects analysis, master load design, wireless individual item mapping, edge-of-failure testing, and F_0 lethality analysis into a single reproducible engineering sequence. It was developed, tested, and applied under real GMP manufacturing conditions at a biopharmaceutical oncology facility supporting an FDA-approved drug product conditions that produced zero-observation regulatory inspection outcomes. The methodology is not theoretical. It has been validated against the most demanding standard available: a federal pre-approval inspection.

The contribution addresses three documented industry problems.

Invisible Failure Modes in Complex Loads

Standard chamber sensors cannot detect sterilization failures caused by residual non-condensable gases or inadequate air removal from long lumens, porous filters, and dense configurations. These failures produce compliant cycle records while actual microbial inactivation is compromised. The pressure-temperature correlation diagnostic developed in this framework grounded in Dalton's law of partial pressures converts air detection from a qualitative judgment into a calculable engineering measurement. No equivalent diagnostic tool exists in current international guidance documents.

The Gap Between Regulatory Expectation and Audit-Ready Evidence

FDA and EMA Pre-Approval Inspections require manufacturers to demonstrate not just that a cycle passed, but that it was designed through a mechanistic, risk-based process. Facilities without a structured engineering methodology cannot produce this record. The framework in this manual generates precisely the traceable, parameter-justified, failure-tested evidentiary package that federal inspectors now treat as a baseline expectation a standard the author helped define through direct PAI support experience.

Lack of a Transferable, Equipment-Agnostic Validation Methodology

Existing industry practice requires each facility to redevelop autoclave validation methodology independently, without a defensible engineering template. The framework presented here is equipment-agnostic and facility-independent. Its application is not limited to oncology manufacturing. Any GMP biopharmaceutical site developing or modernizing a contamination control strategy can implement the same engineering sequence without rebuilding the scientific foundation from first principles.

Steam sterilization has been treated as a testing exercise. This manual reframes it as engineering design and provides the tools, calculations, and documented evidence to defend that position in front of a regulatory authority.

Conclusion

In this methodology, it was proposed a comprehensive, science-based engineering framework that advances steam sterilization cycle development from a fragmented validation exercise to a disciplined, end-to-end methodology grounded in thermodynamic understanding and risk-based design. This no-compromise approach, commonly called Zero-to-Validated, addresses all of the life cycle phases of a sterilization process from blank-sheet design to final defense before regulatory or customer auditors Data-rich studies of the complex thermodynamic behavior show that the success or failure of sterilization can be traced to the degree of non-condensing thermally insulating gas removal and moisture-saturated steam delivery to the pre-vacuum conditioning phase, and not just the duration of the high-temperature exposure The deep modification of the FMEA, the forcing of an extreme master load and the wireless thermal mapping of temperature-discriminating devices create a multilayered barrier to protecting against undetected sterilization failures

Execution of rigorous destructive edge-of-failure studies elevates process verification to a qualitatively new evolutionary level. This practice enables not only the documentation of a static successful test result but also the mathematical substantiation of a large thermodynamic safety margin against unforeseen deviations in production. Precision F_0 lethality calculations, inseparably linked to stringent P/T correlation control and millisecond-level monitoring of equilibration time, ensure attainment of a sterility assurance level not worse than 10^{-6} under any realistically conceivable manufacturing conditions.

From a forward-looking perspective, this risk-based methodology exhibits universal operational scalability. It is applicable to any modern biopharmaceutical facility seeking to implement fail-safe, fully audit-ready sterilization strategies for increasingly complex product and equipment geometries. Under the continuously tightening oversight of international regulatory agencies, the ability of an engineering team to provide scientific justification for every parameter ceases to be a competitive advantage and becomes a primary prerequisite for survival and compliance. The enduring value of this work lies in facilitating a global conceptual transition of the biopharmaceutical industry from an outdated reactive paradigm of Validation through Testing toward an advanced proactive doctrine of Validation through Design. The instrument substantially elevates the operational reliability of complex manufacturing systems, reduces the frequency of unexplained deviations to negligible levels, and ultimately guarantees the highest, uncompromised level of patient safety worldwide.

References

1. Baker, J., Rana, Y. S., Chen, L., Beary, M. A., Balasubramaniam, V. M., & Snyder, A. B. (2025). Superheated steam can rapidly inactivate bacteria, but manual operation of commercial units resulted in limited efficacy during dry surface sanitization. *Journal of Food Protection*, *88*(3), 100461. https://doi.org/10.1016/j.jfp.2025.100461

2. Ben-David, A. (2022). Flexible loading pattern approach in overkill steam sterilization based on the physical properties of steam and thermodynamics of sterilization. *PDA Journal of Pharmaceutical Science and Technology*, *77*(3), 166–180. https://doi.org/10.5731/pdajpst.2021.012708

3. Elmadhoun, B., Alsaidalani, R., & Burczynski, F. (2025). Quality Risk Management in the Final Operational Stage of Sterile Pharmaceutical Manufacturing: A Case Study Highlighting the Management of Sustainable Related Risks in Product Sterilization, Inspection, Labeling, Packaging, and Storage Processes. *Sustainability*, *17*(4), 1670. https://doi.org/10.3390/su17041670

4. European Commission. (2022). *Annex 1 of Volume 4 EU Guidelines for Good Manufacturing Practice for Medicinal Products for Human and Veterinary Use*. European Commission. https://health.ec.europa.eu/system/files/2022-08/20220825_gmp-an1_en_0.pdf

5. FDA. (2009). *Guidance for Industry Q8(R2) Pharmaceutical Development*. FDA. https://www.fda.gov/media/71535/download

6. Feurhuber, M., Taupitz, T., Mueller, F., Frank, C., Hochenauer, C., & Schwarz, V. (2023). A Method to Investigate Sterilization Processes and the Bacterial Inactivation Resolved in Time and Space. *PDA Journal of Pharmaceutical Science and Technology*, *78*(3), 331–347. https://doi.org/10.5731/pdajpst.2022.012771

7. Galiën, R. van der, Langen, A. L., Jacobs, L. J. M., Hagen, B., Flahive, K., Chatterjee, S. D., & Amsterdam, M. C. van. (2023). Setup of a Contamination Control Strategy Using the Hazard Analysis Critical Control Point (HACCP) Methodology. *PDA Journal of Pharmaceutical Science and Technology*, *77*(4), 317–328. https://doi.org/10.5731/pdajpst.2022.012783

8. ISO. (2007). *ISO 11140-5:2007*. ISO. https://www.iso.org/standard/40141.html

9. ISO. (2024). *ISO 17665:2024*. ISO. https://www.iso.org/ru/standard/80271.html

10. Jain, S., Krishnan, N. J. S., Jayaprakash, J. W., Pillai, R. P., Jaganathan, J. N., & Raju, B. V. (2024). Lifecycle Approach to Process Validation and Its Implementation in Pharmaceutical Industry. *International Journal of Pharmaceutical Quality Assurance*, *15*(3), 1845–1856. https://doi.org/10.25258/ijpqa.15.3.113

11. Miguel, E. A., Patrícia, F., Laranjeira, P. R., Ishii, M., & Andreoli, J. (2024). Steam quality monitoring as a strategy to reduce wet packs and sterilization failure. *American Journal of Infection Control*, *52*(11), 1314–1319. https://doi.org/10.1016/j.ajic.2024.07.002

12. Pletzer, S., Lang, B., Miranda, M., Bergmann, A., & Hochenauer, C. (2025). Modeling steam penetration into hollow devices: Effects of phase change and non-condensable gases during steam sterilization. *International Journal of Heat and Mass Transfer*, *256*, 127933. https://doi.org/10.1016/j.ijheatmasstransfer.2025.127933

13. Pletzer, S., Miranda, M., Lucchesi, M., Magno, M., & Hochenauer, C. (2022). Numerical Investigation of the Heating Process of the Walls in Steam Sterilizers. *Chemie Ingenieur Technik*, *95*(9), 1457–1466. https://doi.org/10.1002/cite.202200135

14. Ramos, I., Najera, M., & Schaefer, G. (2024). Integration of rapid bioburden testing into production quality management systems and process control. *Biotechnology Progress*, *40*(3), e3431. https://doi.org/10.1002/btpr.3431

15. Rutala, W. A., Donskey, C. J., & Weber, D. J. (2023). Disinfection and sterilization: New technologies. *American Journal of Infection Control*, *51*(11), A13–A21. https://doi.org/10.1016/j.ajic.2023.01.004

16. Stewart, A. (2025). Implementation of Continued Process Verification Program for Post-Market Compliance and Product Success. *PDA Journal of Pharmaceutical Science and Technology*, *79*(4), 470–471. https://doi.org/10.5731/pdajpst.2025.25424

17. Tessarolo, F., & Masè, M. (2025). Field assessment of a novel sensor for measuring noncondensable gases in steam sterilizers. *Scientific Reports*, *15*, 9836. https://doi.org/10.1038/s41598-025-94798-1

www.ingramcontent.com/pod-product-compliance
Lightning Source LLC
LaVergne TN
LVHW052150210926
848482LV00019B/335
9798905101595